The summer day is hot and still. The sky gets dark quickly. The world waits for the storm to start.

Soon a cool rain wets the earth. The dry corn drinks it up. Happy ducks quack in the rain.

Kids at play quit their sports. They play word games on the porch. Or they play inside. But other people work in the rain.

It rains quite hard for a long time. The raindrops form little puddles. The little puddles get big.

There is a flash of light.
Thunder cracks and booms!
Rain fills the ponds and lakes.

Cars honk their horns. Ships wait in port. Will this storm be short or long?

At last the rain stops.
The sun shines in the sky.
The world seems bright
and fresh.

Kids play in the cool air.

Soon the puddles will dry up.

The storm has ended.

The End

Understanding the Story

Questions are to be read aloud by a teacher or parent.

1. What is this story about?
2. How does the rain help things?
3. Why is the rain sometimes a problem for children?
4. What is the world like after a storm?

Answers: 1. a storm 2. Possible answers: by watering plants, making ducks happy 3. Possible answer: because they have to play inside 4. bright and fresh

Saxon Publishers, Inc.
Editorial: Barbara Place, Julie Webster, Grey Allman, Elisha Mayer
Production: Angela Johnson, Carrie Brown, Cristi Henderson

Brown Publishing Network, Inc.
Editorial: Marie Brown, Gale Clifford, Maryann Dobeck
Art/Design: Trelawney Goodell, Camille Venti, Andrea Golden
Production: Joseph Hinckley

© Saxon Publishers, Inc., and Lorna Simmons

All rights reserved. No part of the material protected by this copyright may be reproduced or utilized in any form or by any means, in whole or in part, without permission in writing from the copyright owner. Requests for permission should be mailed to: Copyright Permissions, Harcourt Achieve Inc., P.O. Box 27010, Austin, Texas 78755.

Published by Harcourt Achieve Inc.

Saxon is a trademark of Harcourt Achieve Inc.

Printed in the United States of America
ISBN: 1-56577-988-6

5 6 7 8 546 12 11 10 09 08

Phonetic Concepts Practiced

or (storm, or)
qu (quickly)

Nondecodable Sight Words Introduced

word
work
world

ISBN 1-56577-988-6

Grade 1, Decodable Reader 26
First used in Lesson 74

Andrew

Fox, Not Ox!

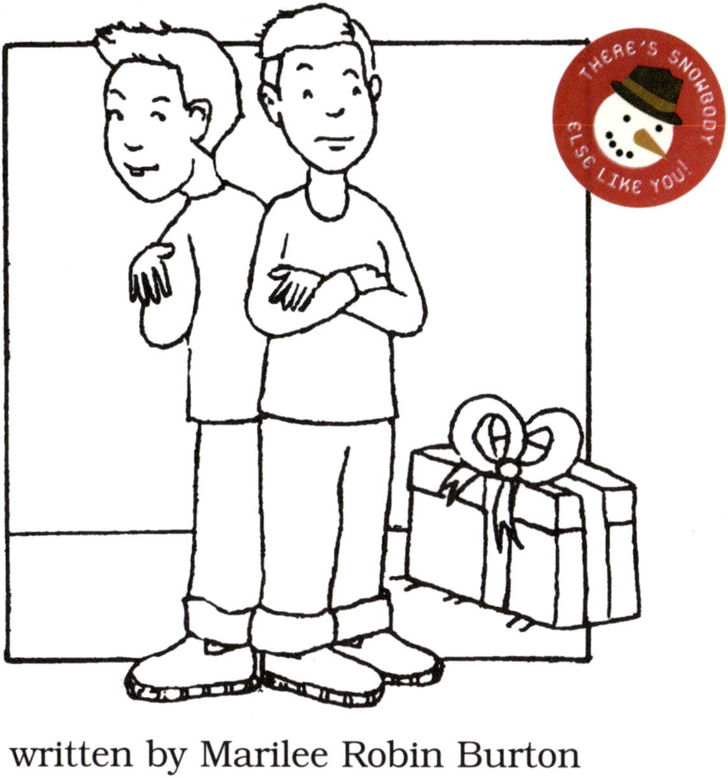

written by Marilee Robin Burton
illustrated by Maggie Byer-Sprinzeles

SAXON PUBLISHERS

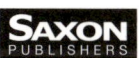

THIS BOOK IS THE PROPERTY OF:

STATE_____
PROVINCE_____
COUNTY_____
PARISH_____
SCHOOL DISTRICT_____
OTHER_____

Book No. _____
Enter information
in spaces
to the left as
instructed

| ISSUED TO | Year Used | CONDITION ||
		ISSUED	RETURNED

PUPILS to whom this textbook is issued must not write on any page or mark any part of it in any way, consumable textbooks excepted.

1. Teachers should see that the pupil's name is clearly written in ink in the spaces above in every book issued.
2. The following terms should be used in recording the condition of the book: New; Good; Fair; Poor; Bad.